To our readers, those students just beginning a life of adventure and learning, always remember:

You are braver than you believe, stronger than you seem, and smarter than you think.

~

A.A. Milne

Adapted from the poem *The Months* by Sara Coleridge, who lived in England from December 23, 1802 to May 3, 1852.

Created by

Brent A. Ford

© 2017 by nVizn Ideas LLC

www.nviznideas.com

January brings the snow,

Makes our feet and fingers glow.

February brings the rain,

Thaws the frozen lake again.

March brings breezes loud and shrill,

Stirs the dancing daffodil.

May brings flocks of pretty lambs,

Skipping by their fleecy dams.

June brings tulips, lilies, roses,

Fills the children's hands with posies.

Hot July brings cooling showers,

Apricots and gillyflowers.

August brings ears of corn,

Then the harvest table is born.

Warm September brings the fruit,

Bears then begin to loot.

Fresh October brings the pheasant,

Then to gather nuts is pleasant.

Dull November brings the blast,

Then the leaves are whirling fast.

Chill December brings the sleet,

Blazing fire and holiday treat.

Ozzie & Alina Adventures

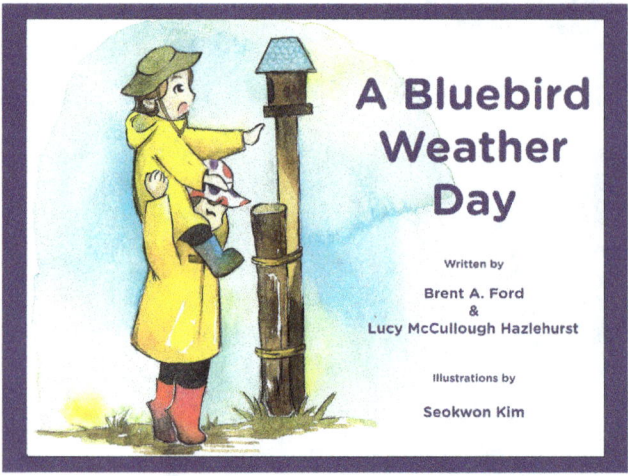

Updated Classics

Adventures from nVizn Ideas

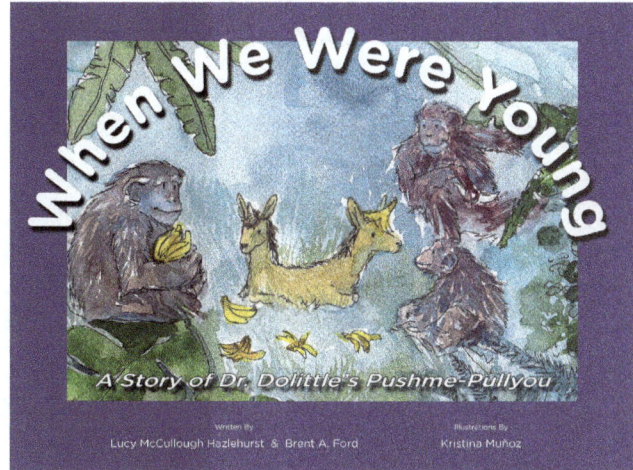
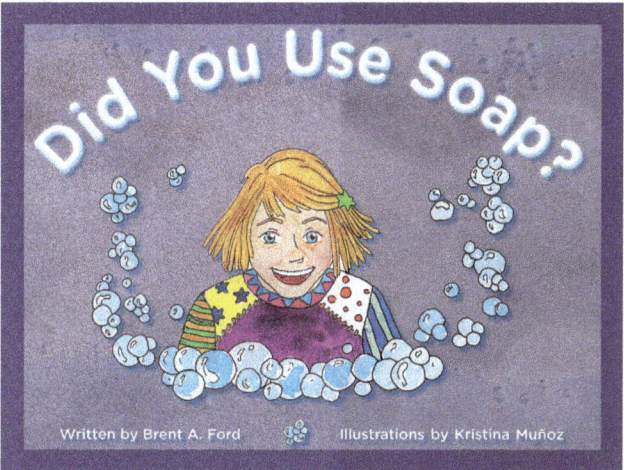

Science & Nature eBooks

Interactive eBook

Variety in the Animal World

Well, Will You Look at That?

Well, Will You Look at That?

Well, Will You Look at That?

Well, Will You Look at That?

Well, Will You Look at That?

Well, Will You Look at That?

www.ingramcontent.com/pod-product-compliance
Lightning Source LLC
Chambersburg PA
CBHW080811040426
42333CB00062B/2706